GENETICS

ROBERT SNEDDEN

Wayland

CONTENTS

First published in 1995 by
Wayland (Publishers) Limited
61 Western Road, Hove, East
Sussex BN3 1JD, England
© Copyright 1995 Wayland
(Publishers) Limited

British Library Cataloguing in
Publication Data.
Snedden, Robert
Genetics: Advances That Have
Changed the World. – (Science
Discovery Series; Vol.5)
I. Title II. Burns, Robert III.
Series 613
ISBN 0-7502-1236-5

Acknowledgements
Concept David Jefferis
Text editor Michael Brown
Diagrams Robert and Rhoda
Burns/Drawing Attention
Line illustrations James Robins

Picture credits
Bruce Coleman Ltd 7 (TL: Erwin
and Peggy Bauer, TR: Jean-
Pierre Zwaenepoel), 9 (Len Rue
Jr), 11 (Dieter and Mary Plage),
12 (Jen and Des Bartlett), 14
(Frans Lanting, 46 (Jane Burton)
Delta Archive 13, 16, 18, 19,
20(TL), 22, 23(BR), 33, 41, 45

Mary Evans Picture Library 15
Science Photo Library 4, 5,
20(B), 23(T), 24, 25, 26, 28(B),
29, 30(B), 31, 32, 34(TL), 39, 40

Printed and bound in Italy by
G. Canale and C.S.p.A., Turin

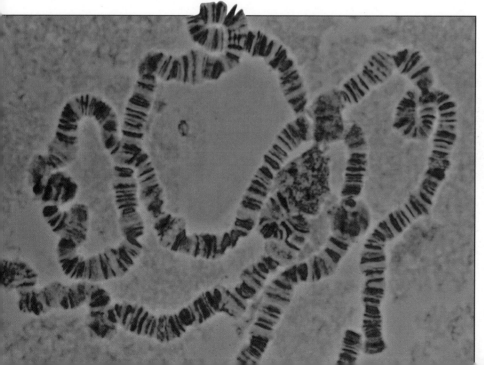

INTRODUCTION

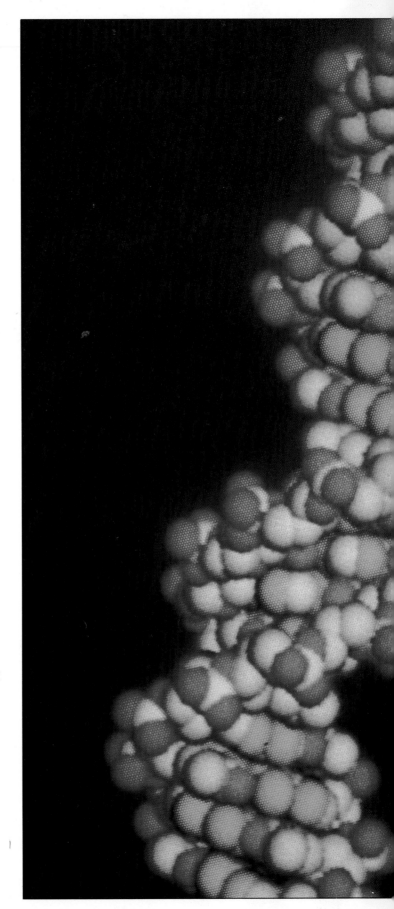

Almost everyone has heard of dinosaurs, the great reptiles that dominated the Earth over 65 million years ago. The fact that these beasts and many other types of plants and animals that once were common no longer exist shows that life has been changing since it first appeared. The first simple forms of life that appeared around 3,500 million years ago have somehow given rise to the millions of different kinds of living things that share our world today.

Yet at the same time these creatures and plants hardly seem to change at all. Female dogs give birth to puppies, not kittens; chickens hatch from chicken eggs, not duck eggs and if you plant pea seeds you do not expect roses to grow.

For thousands of years farmers have known that animals and plants could pass on many characteristics to their offspring. Sheep that gave particularly good wool could be bred from and their lambs would tend to have good wool too. Plants could be crossbred to produce varieties with desirable features, such as juicy fruit or plentiful seeds. But just how were these features passed on? What is it that living things have that allows new forms to appear over millions of years, but that still keeps the offspring looking more or less like their parents? It is the answer to these sorts of questions that we will be looking for in this book.

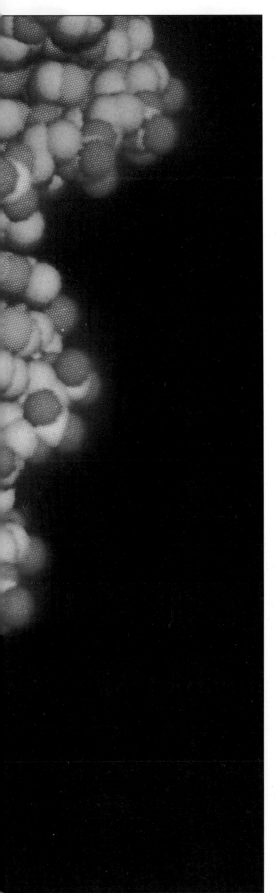

 # WHAT IS GENETICS?

Genetics is one of the youngest of the sciences. It is the branch of biology that is concerned with the study of heredity, the passing on of characteristics from one generation to the next, and with the variations that distinguish one living thing from another. The beginning of genetics can be pinpointed almost exactly, to the work of an Austrian monk called Gregor Mendel who put forward a theory of heredity in 1865 based on his observations and experiments growing sweet peas in a monastery garden. Mendel's work provided a starting point from which others set out to look for the part of each living thing that was responsible for determining its characteristics and for generating new variations. It was a search that was to end less than seventy years after Mendel's death with the discovery of a remarkable molecule – DNA.

DNA is at the heart of every living cell and is the blueprint, or masterplan, for all life on Earth. Parts of the DNA called genes control the way an organism grows and develops. Using a simple genetic code, DNA has produced the bewildering variety of lifeforms that share the Earth with us. All living things have their own unique DNA code, but the molecule that forms that code is fundamentally the same. Over the past thirty years, our understanding of genetics has transformed research into biology, medicine and drugs. Most scientists are using genetic techniques to understand living processes and bring about dramatic improvements in people's health.

◀ **The master blueprint for life itself, the DNA molecule, that winds in the form of a spiralling double helix.**

SPECIES – FIXED OR CHANGING?

Long before Mendel and his study of heredity, biologists began to ask themselves how, and if, species ever changed or developed. They were especially influenced by the work of the great Swedish botanist Carl Linnaeus (1707-1778). In 1735, he published a book called *Systema Naturae* (System of Nature) in which he described individual types, or species, of plants and animals and showed how they differed. Species with similar features he combined together in groups called genera. Closely related genera were further combined as orders, and orders combined to give classes.

Linnaeus believed that the natural order he had uncovered was the work of God and that all living things in the world had been created at the same time and their forms were fixed and unchanging. Later, he changed his views a little and suggested that God had created only one species in each genus, and that the variety we see today came about as these original species were crossed and their characteristics seen in different combinations. Along with many other naturalists of the seventeenth and eighteenth centuries, Linnaeus found it hard to believe that anything could ever become extinct because this would leave a gap in God's creation.

▲ The work of Linnaeus covered a total of almost 12,400 plant and animal species, practically all the creatures then known.

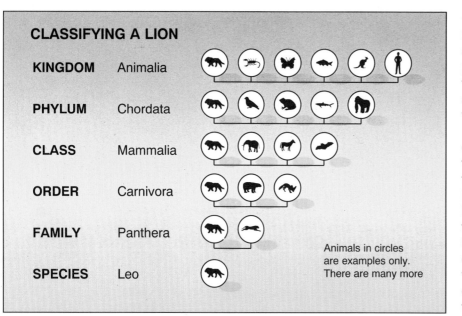

CLASSIFYING A LION

KINGDOM	Animalia	
PHYLUM	Chordata	
CLASS	Mammalia	
ORDER	Carnivora	
FAMILY	Panthera	
SPECIES	Leo	

Animals in circles are examples only. There are many more

◀ Living things are classified into the wide-ranging plant and animal kingdoms, then through progressively tighter groupings to species, creatures that are so alike that they can only breed within that group.

In the eighteenth century no one knew why living things breed true, and among the theories was that of Swiss naturalist Charles Bonnet (1720-1793), who believed in 'preformation'. This claimed that every creature is preformed in the egg, with yet another, tinier descendant within it, and so on, without end. This was thought to explain why creatures give birth to similar descendants.

▲ The lion and tiger are both members of the family *Panthera*. To distinguish these species the lion is known as *Panthera leo*, the tiger is *Panthera tigris*.

By the middle of the eighteenth century, the new science of geology was beginning to reveal that the Earth had changed over millions of years and was still changing. If the Earth itself had changed then, surely, life on Earth would have to adapt and change, too? The fossils that were being uncovered seemed to suggest that some creatures hadn't changed and had died out as a consequence.

It was the French naturalist Georges Buffon (1707-1788) who introduced more modern ideas about the development of species. He believed that similar species, such as the horse and the donkey, had a common ancestor and that a member of one species cannot breed successfully outside the group. However, he also said that what Linnaeus called a genus was simply a different form of the same species. For example, there was a horse species, which included donkeys and asses, and a cat species, which included tigers and pumas. Buffon believed that changes appeared in a species if it was forced to live in a new environment, with the result that the species took on a new form. These changes were not necessarily for the better – Buffon said that the donkey was a horse that had degenerated through having to live in an inferior environment. Linnaeus disapproved of Buffon and named a foul-smelling plant after him!

☀ BINOMIAL CLASSIFICATION

Linnaeus's lasting contribution to biology was his system of giving a two-part (binomial) Latin name to every species. This system is still used. For example, *Canis lupus* is the scientific name for the animal commonly known as the wolf. The first part of the name, 'Canis', shows that the animal belongs to the dog family.

ADAPTING AND EVOLVING

▲ French biologist Jean Baptiste Lamarck was one the first to think up a reasoned theory of evolution. Later research showed there were basic errors in his thinking, but his ideas were taken up by Trofim Lysenko (1898-1976) in Soviet Russia. However, Lamarckian theories did not stand up in practice and Lysenko's work was disproved.

The idea of gradual change introduced by Buffon was discussed further in a book called *Zoonomia*, written by Erasmus Darwin (1731-1802) and published between 1794 and 1796. Darwin put forward the idea that living things adapted to changes in their surroundings. These adaptations could then be passed on to their offspring. Each generation built on and added to the changes that had occurred in previous generations. Darwin's ideas were similar to those put forward by the French naturalist Jean Baptiste Lamarck (1744-1829) a few years later. Lamarck was the first person to come up with a properly thought-out explanation for the way life might have developed from one form to another. He believed that the variety of living things we see today had come about through the first simple organisms changing gradually over many generations as they adapted to their environment.

Lamarck said that the characteristics an animal or plant acquired during its lifetime could be passed on to its offspring, a process known as 'the inheritance of acquired characteristics'. This was a bit like saying that because a tennis player develops one arm bigger and stronger than the other as a result of continual use on the tennis court, his or her children would have one, too.

As an example of his theories in practice, Lamarck pointed to the giraffe. Imagine, he said, that once upon a time there was a type of antelope that liked to eat tree leaves. These antelopes would stretch upwards to nibble as many leaves as possible, including those at the tips of the upper tree branches. In the course of the antelopes' lives all this stretching would make their necks and legs a little bit longer. According to Lamarck, the antelopes' offspring would then be born with slightly longer necks than those that their parents had when they were born. Bit by bit, for generation after generation, the antelopes would develop longer necks and legs. As a result, eventually the antelopes would have become a new species – the giraffe.

◄ Lamarck's idea that animals could improve themselves physically – and pass this on as a genetic improvement – turned out to be mistaken. As you will see later, reality seems to be that creatures do adapt to a changing environment, but that the changes are random and modified by natural selection. Just because this giraffe learns to stretch to the highest branches does not mean that its offspring can do the same.

☀ ENOUGH TIME FOR LIFE TO EVOLVE?

For a long time people's religious beliefs got in the way of any serious thinking about evolution. Many people thought that the Earth was only about 6,000 years old, which would not leave enough time for major evolutionary changes to take place.

The Scottish geologist James Hutton (1726-1797) was one of the first scientists to suggest that the surface features of the Earth had been formed by processes taking place over not thousands, but many millions of years. This would give enough time for slow changes to take place in lifeforms as well.

Although there were flaws in his theory, Lamarck did a great deal to win acceptance for the idea that species could change and evolve, and for that reason his work was of great importance. However, like Linnaeus, Lamarck would not accept that species could become extinct. Fossils were evidence that life had changed, not that it had disappeared. Thirty years after Lamarck's death, Erasmus Darwin's grandson was to publish a book that not only disproved their ideas, but also showed how species could be driven to extinction by changes in their environment.

CHARLES DARWIN

Erasmus Darwin's grandson Charles was born in Shrewsbury, England, in 1809. The son of a doctor, Charles tried to follow in his father's footsteps and went to Edinburgh University in 1825 to study medicine, but could not stand the sight of blood or the horror of operations conducted in unhygienic conditions without anaesthetics. In 1827, he went to Cambridge, intending to join the Church, but while there he turned more and more from his studies to his great love – natural history.

Four years later Darwin's life was to change dramatically. Through contacts he made with scientists at Cambridge, he was offered the position of ship's naturalist on board HMS *Beagle*, which the British Admiralty had commissioned to survey the coast of South America. Darwin accepted eagerly, and in December 1831, the *Beagle* set sail from Devonport on a five-year voyage that would take it round the world. It was perhaps the most important journey ever made in the history of biology.

▲ Charles Darwin was ship's naturalist aboard HMS *Beagle* from 1832 to 1836. He made extensive notes on everything that he saw on the journey.

▲ The Galapagos Islands are situated in splendid isolation, conditions that have allowed unique lifeforms to develop.

During his travels Darwin uncovered a multitude of fossils on the South American mainland and saw how some species became extinct while others survived. This seemed to him to disprove the still-popular notion that life was changed by terrible catastrophes. He also visited the Galapagos Islands in the Pacific Ocean, about 1,000 kilometres off the coast of Ecuador. Here, he gathered specimens from a group of birds that have become known as Darwin's finches. There are 13 different species of finch on the Galapagos Islands, each one confined to its own particular island. All of the finches are very similar to each other, differing only in the size and shape of their beaks, and they are all unique to the Galapagos. Not one of them is found anywhere else. It seemed puzzling to Darwin why these birds should have appeared only on these tiny islands.

▲ Apart from
finches, the
Galapagos
Islands offer
other unique or
rare species,
such as this
land iguana.
The bird on top
is a type of
finch that takes
dead skin and
insects from
the iguana's
body!

It was some time later – after the voyage was over, in fact – that Darwin came up with a solution. He believed that the finches all shared a common ancestor – a single species of finch that had found its way to the islands from the mainland thousands of years earlier. On the islands, the finches had no competition from other birds and they had changed and evolved to fit the range of different conditions found on the islands. Some had adapted to eat insects and had small sharp beaks, those that ate seeds had large, heavy beaks and one, the woodpecker finch, had even learned to use a cactus spine held in its beak to probe for insects beneath tree bark.

However, Darwin did not know how the finches had changed, even if he had guessed why. He was not happy with Lamarck's idea of gradual change through acquired characteristics being passed from generation to generation. He began to look for answer by studying the ways in which humans produce new breeds of plants and animals.

NATURAL SELECTION

▲ According to Thomas Malthus, feeding the poor encouraged them to have more children, so multiplying their problems. Improving standards of living seems to improve things however – in the 1990s, rich countries have lower birth rates than poor countries.

For centuries, farmers have carefully selected for breeding those animals and plants that have the qualities they want to maintain. Only the corn that produces the most seeds is planted in the spring; only the cows that give the most milk will have calves. It was this idea of selection to produce the best plant or animal for a purpose that gave Darwin part of the inspiration he was searching for.

Another push forward came in September 1838, when Darwin read a book called *An Essay on the Principle of Population*, written by Thomas Malthus (1766-1834), a clergyman and political economist. In this book, published in 1798, Malthus argued that human populations inevitably grew too big to be sustained by their supplies of food. When this happened, the strongest members of the society would have the best chance of surviving, while the weaker members would suffer through starvation, war and disease. Darwin had the idea that this argument could apply to other forms of life as well. Plants and animals had no way of increasing their food supply and so those species that could make the most efficient use of existing food would have an advantage. He suddenly saw all species as being caught up in what he later called, 'the struggle for existence'.

◀ Hunter and hunted sum up the struggle for existence in the basic urge, to eat. The prize for the winner is survival and the chance to reproduce. Here a lion runs at zebra and springbok.

◀ Friesian cows are an example of a breed created by artificial selection. In this case, high milk yield was the main aim.

Different characteristics would be vitally important in determining whether or not a creature would survive. It might be speed or strength, or, in the case of the Galapagos finches, having the right shape of beak to eat the food available. Those that were best adapted to their surroundings would have the best chance of surviving long enough to breed and have offspring. To take Lamarck's example, the antelopes didn't get longer necks by stretching. Some of them were born with longer necks and they had an advantage over the others because they could get more food. This meant that they lived longer and had more offspring. In contrast, short-necked antelopes might possibly starve when food was in short supply because they couldn't reach as much. Over time, all the antelopes would have long necks.

In the same way that farmers produced new breeds through artificial selection, so, Darwin said, new species appeared in nature through natural selection. The biologist Herbert Spencer (1820-1903) used the phrase 'the survival of the fittest' and Darwin agreed that this was a good way to describe the process. After his return from the *Beagle* voyage, Darwin wrote several books setting out his discoveries. The first one he published, in 1839, was called *A Naturalist's Voyage on the Beagle*. It was a very popular book and made him famous – it is still available today.

☀ SURVIVAL OF THE FITTEST

In 1977, a drought on the Galapagos Islands brought an opportunity to see natural selection at work among Darwin's finches. The drought wiped out most of the supply of small seeds on the islands leaving only large seeds. Birds with small bills could not cope with the larger seeds and were soon replaced by those with bigger bills which could eat the seeds. Big-billed birds had always been there, but the drought gave them the opportunity to multiply in higher numbers than before.

THE ORIGIN OF SPECIES

▲ **Down House, Darwin's home in Kent, is now a museum with many of Darwin's items on view to visitors. Here some stuffed Galapagos finches are laid out next to an entry in Darwin's 1835 field journal and a first edition of *Origin of Species*.**

Darwin first put his ideas on evolution down on paper in 1842, four years after reading Malthus, but did not show them to anyone or attempt to have them published. Perhaps he was afraid of how people might react. For the next fourteen years Darwin continued to think about evolution and to gather information that would help him to prepare a strong case. During this time he continued to work on and write about other aspects of natural history.

In 1856, Darwin at last began working in earnest on his book on natural selection. Two years into the work, in June 1858, he received a package at his home, Down House in Kent, that amazed him. In it was an essay written by the naturalist Alfred Russel Wallace (1823-1913), which contained a brief and accurate summary of the ideas on evolution that Darwin had been considering for almost 20 years. Wallace had spent ten years travelling around South America and Malaysia studying and collecting information for his particular interest, the origin of species. He had also read Malthus, and he too, had become convinced that evolution took place through natural selection.

▼ **Darwin collected an enormous number of beetles when he was a student at Cambridge University. In fact, he was there to study to be a clergyman, but instead developed an enthusiasm for natural history.**

The two men agreed to work together and a joint presentation of their ideas was made to the learned Linnaean Society in London on 1 July 1858. Darwin was quite prepared to allow Wallace the credit of being first to publish a theory of evolution by natural selection, but Wallace was a very modest man who never insisted on the recognition he deserved. In fact, years later in 1889, after Darwin's death, Wallace published *Darwinism*, a book examining various aspects of Darwin's theories. It is a great shame that we speak of 'Darwin's theory of evolution' while the work of Wallace is mostly forgotten.

Soon after he received Wallace's essay, Darwin decided to write a shorter version of the massive book he had been working on since 1856. In 1859 it was published as *On the Origin of Species by Means of Natural Selection*, usually shortened to *The Origin of Species*. It is one of the most famous books ever written, and at the time caused great uproar. For many people the idea of evolution threatened their religious beliefs – they thought that humans were special. But Darwin showed that we, and all other species, had evolved from earlier forms of life. In fact, what really upset many people was Darwin's belief that humans and the great apes share an ancestor.

There were still two questions that natural selection could not answer, however. How did variations come about in living things in the first place? And how were they passed on from parents to their offspring? Completely unknown to Darwin, an Austrian monk called Gregor Mendel was finding some answers to the second question. The answer to the first question lay a hundred years in the future.

☀ DARWIN'S BULLDOG

Thomas Henry Huxley (1825-1895) was a distinguished zoologist and supporter of Darwin's theory of evolution. So willing was he to engage important clergymen in fierce debate about the theory that he earned the title 'Darwin's Bulldog'. In one famous instance Samuel Wilberforce, Bishop of Oxford, asked Huxley if he was descended from an ape on his mother or his father's side. Huxley's reply was said to be so devastating that a lady in the audience fainted and had to be carried out!

◀ **Darwin's ideas exploded across Victorian society like a bomb. Many people were upset and some poked fun, as this 1882 cartoon from *Punch* magazine reveals.**

THE BIRTH OF GENETICS

▲ Gregor Mendel's enthusiasm for scientific study was shared by several others at his monastery. Although his work went largely unnoticed until the twentieth century, it laid the foundations for the whole science of genetics.

Gregor Mendel (1822-1884) was born in Heinzendorf, Austrian Silesia (now part of the Czech Republic). As a child he helped his father to tend fruit trees. In 1843, he entered a monastery, the Abbey of St Thomas, at the town of Brünn (now called Brno) and became a priest in 1847. While at the monastery, the abbot in charge sent Mendel to the University of Vienna, where he studied physics, chemistry, zoology, botany and mathematics, but he failed to pass the examination to qualify as a teacher. After his return to Brünn, Mendel used his skills as a gardener and his interest in evolution in an eight-year series of experiments on pea plants in the monastery garden. It was in the course of these experiments that he laid the foundations of the modern science of genetics.

Mendel was interested in discovering how different features were passed from parent plants to their offspring. The seven characteristics he chose to study came in one or other of two forms. They included, for example, the height of the plant, which could be either tall or short, the colour of its seeds (either yellow or green) and the shape of the seeds (either smooth or wrinkled).

◄ These plants may look similar, but as researchers have shown, they are all individuals, just as no two humans are exactly the same.

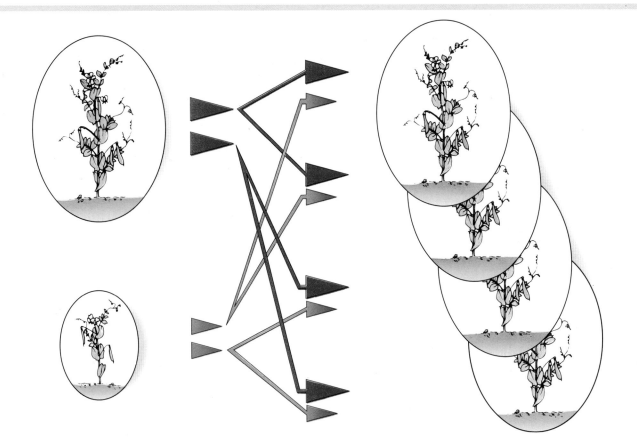

Mendel took plants that always produced one form of a characteristic, say tall plants, and crossed them with plants that always produced the opposite form, short plants in this case. He did this by transferring pollen from the flowers of the tall plants to those of the short plants. Then he planted the seeds that the plants produced to see how the offspring – called the first filial generation – would turn out. He discovered that they were not a mixture of tall and short plants, as might have been expected, nor were they all medium-height plants. In fact, the hybrids – the name given to the offspring of parents that differ in one or more characteristics – were all tall.

Mendel then allowed these tall hybrids to self-pollinate and planted their seeds to see what the second generation would produce. This time he found that three-quarters of the plants were tall and the rest were short, a ratio of three to one. Mendel crossed hundreds of plants and kept careful records of his findings. The same thing happened with the other characteristics of the pea plant that he investigated. The second generation of plants always showed the alternative forms of the characteristic he was investigating in a three to one ratio. What could have been the explanation for this?

▲ Mendel used pea plants for his experiments. He found patterns in sizes and colours that reached through the generations. This meant that Mendel was able to come up with laws that could predict the appearance of plants, depending on the qualities of their parents.

In this diagram, tall plants (red triangles) have a dominant gene. When crossed with short plants (blue triangles) the offspring are all tall. In the second generation however, one in four will be a short plant.

MENDEL'S LAWS

In seeking an answer to how a plant's characteristics were passed on to the next generation, Mendel decided that his plants must contain 'particles of inheritance', what we know today as genes. It was these 'particles' that determined how the plant would appear. Each plant, he believed, had a pair of particles for each of its characteristics, one from each parent. There was a particle for 'tall', one for 'short'; one for 'green seeds', another for 'yellow seeds', and so on.

To explain the three-to-one ratio he had discovered, Mendel proposed that one particle for a particular characteristic dominated that for the opposite characteristic. For instance, in the parental generation, tall plants crossed with short ones gave only tall plants; green-seeded plants crossed with yellow-seeded plants produced plants with only green seeds. He called the tall and green seed factors 'dominant', while short and yellow seeds factors were 'recessive', meaning that they receded or faded into the background.

However, the 'short' and 'yellow seeds' characteristics didn't disappear. In every case, crossing the first generation plants with themselves produced both tall/short and green/yellow features in a three-to-one ratio (for example, three tall plants and one short one). All seven of the characteristics Mendel investigated in the pea plant produced the same results.

▲▼ It is important to recognize that there is a difference between environmental factors and what may be genetically possible.

An elm tree (above) may grow some 30 metres or more, given the right conditions.

Deliberate stunting of a young elm, such as the bonsai shown below, may result in a tree that reaches barely 10 cm tall. Despite this, the bonsai elm's seeds retain exactly the same potential for growth as a normal tree.

From his findings, Mendel suggested that there were laws ruling the way particles were passed down the generations. Two particles must exist for each characteristic. When the plant's sex cells were formed, the pairs of particles separated. Each pollen grain (male) and ovule (female) in a flower, or each egg and sperm in animals, had only one particle for each characteristic. When fertilization took place, each parent would have given one particle per characteristic to its offspring. Which particular pollen grain and ovule, or egg and sperm, joined together was not affected by the particles they carried. Mendel also said that the various particles for different characteristics did not interfere with each other. The colour of a plant's flowers, for example, did not affect whether it was short or tall.

Unfortunately, the flying start Mendel gave to the science that would come to be called genetics was almost completely ignored for decades. He read an account of his work at the Brünn Natural Scientific Society in 1865, and published it in the society's journal. However, no one attempted to follow up his experiments and the German botanist Karl von Nägeli (1817-1891), to whom Mendel wrote, dismissed his work as unimportant. It was only in 1900, sixteen years after his death, that Mendel's paper was rediscovered, and he got the credit he deserved.

▲ Mendel's laws have been used by generations of plant growers to improve existing varieties and to create new ones or mixed, 'hybrid' varieties. In the world of flower growing, luxuriant blooms with rich, bright colours tend to be the most sought after.

✹ POLYGENES

Not all characteristics can be explained simply in terms of a pair of inherited factors. These characters, such as weight and skin colour, are determined by several genes, or polygenes, acting together. The individual genes follow Mendel's laws, but their effects cannot be studied separately. Mendel was lucky because the characteristics he looked at were determined by single genes.

DISCOVERING THE CELL

▲ One of Hooke's microscopes, here shown in a nineteenth century scientific engraving.

▼ Hooke studied widely and was the first to see that plants were made of cells. Today's equipment reveals structures in intricate detail, with magnifications of up to a million times available.

Mendel had no idea what his 'particles of inheritance' actually were. He had drawn up his rules by thinking carefully about the results of his experiments. He knew the particles must be held in the sex cells – the sperm and eggs of animals, the pollen and ovules of plants – but no more than that. To begin the search for the particles we have to go back in time to the invention of the microscope and the discovery of the cell.

Robert Hooke (1635-1703) was a great English scientist. He was one of the first to make serious studies using the microscope, invented in 1608 by the Dutchman Zacharias Jansen (1588-1630). Hooke wrote the very first book of experiments and observations using microscopes. It was called *Micrographia* and was published in 1665. In it, he described how he had examined a very thin slice of cork through a microscope. He saw that it was made up of a fine pattern of tiny regular holes: 'a great many little boxes' is how he described it. He called each little box a 'cell', a word that meant 'compartment', because together they resembled the structure of a honeycomb – the mass of cells in which bees store honey. Hooke's use of the word 'cell' referred to the empty walls of dead plant cells (the cork), but it was later used to describe living cells.

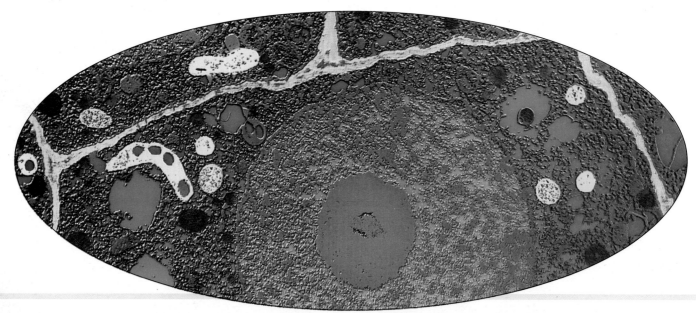

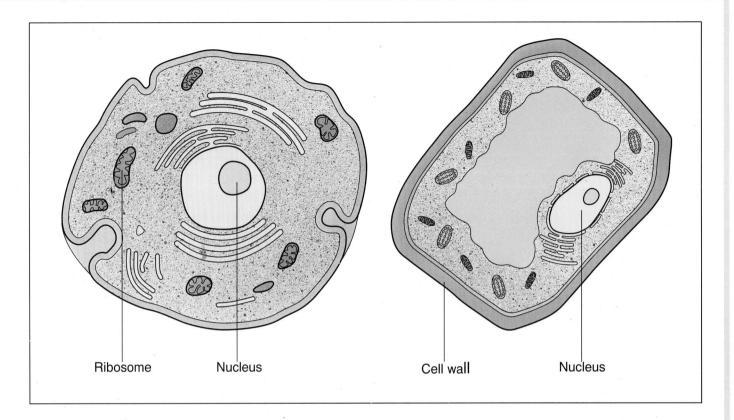

| Ribosome | Nucleus | | Cell wall | Nucleus |

The discovery that living things were made up of cells was very important. Soon, scientists began to investigate further. In 1831, the Scottish botanist Robert Brown (1773-1858) observed that all plant cells had a small object inside them. He called this the nucleus, from a Latin word meaning 'little nut'. The nucleus plays a major part in genetics and we will return to it soon.

Three German scientists played a major part in furthering knowledge about the cell. In 1838, the botanist Matthias Schleiden (1804-1881) made a microscopic study of different types of plant and saw that they were all made up of cells. He thought that new cells appeared from the nuclei, but this idea was later proved wrong. Theodor Schwann (1810-1882), came to the conclusion that animals were made up of cells, too. In 1839, he put forward the 'cell theory', stating that all living things consisted of cells or the products of cells. He also said that each cell could be considered as an individual living thing. Human beings and other complex organisms consisted of billions of cells all working together. Between 1846 and 1848, Carl von Siebold (1804-1885) showed that at the other end of the scale, some microscopic living things were made up of just a single cell. He called this new group of organisms the *Protozoa*.

▲ **Typical animal (left) and plant cells, shown in cross-section. Shape and size varies widely, depending on the function of a particular cell. In your body, there are muscle, nerve and skin cells, amongst many others.**

☀ ELECTRON MICROSCOPE

Studies of cell structure took a leap forward when the electron microscope was developed. The first operational one was built in the 1930s by German scientists in Berlin, and within a few years electron microscopes were providing magnifications 50 times greater than light microscopes. Today's instruments are better still, and can show objects just 100,000th of a mm across.

SPONTANEOUS GENERATION

For centuries, people had argued about the origins of life. Where did living things come from? It was widely believed that under certain circumstances life could arise automatically, or spontaneously, from non-living materials. The Greek philosopher Aristotle (384-322 BC) thought that rotting meat gave rise to flies, for example. Although later scientists, such as the Italian Francesco Redi (1626-1697), showed that this belief was false, many still believed in spontaneous generation. While no one believed that flies and larger animals appeared in this way there were still many – George Buffon, who we met earlier, was one of them – who said that this was how the very smallest living things came into existence. Buffon believed that larger creatures had appeared spontaneously early in Earth's history when, he believed, conditions had been much hotter than they are now.

▶ Buffon was right about one thing – in the conditions that existed at the dawn of life, the Earth was far hotter than it is now. This picture shows a fiery world, the air a mixture of gases and water vapour. Lightning flashes provide the energy that may have helped create the first building blocks of life.

In 1855, Rudolph Virchow (1821-1902), who had been a fellow student with Theodor Schwann, stated that: 'All cells arise from cells.' Virchow was not the first to make this claim, but he was important because he had studied diseased tissue and had shown that healthy cells changed slowly into diseased ones – the sick cells didn't just appear from nowhere. Virchow's research helped bring an end to the idea of spontaneous generation.

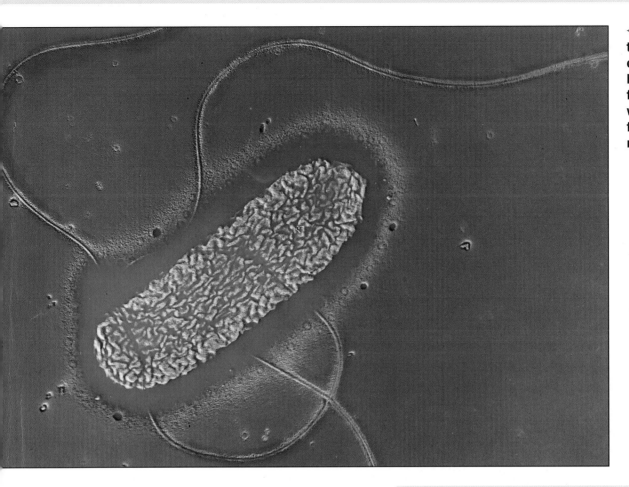

◀ Bacteria are the simplest cells. They have such features as cell walls and flagella for movement.

At about the same time, Louis Pasteur (1822-1895), a French chemist, carried out a series of simple experiments, proving once and for all that spontaneous generation was wrong. He demonstrated that food went off because it became contaminatcd by tiny living organisms carried on dust particles in the air. By keeping the organisms out he was able to stop meat from spoiling.

Pasteur also developed what he called his 'germ theory'. He believed that diseases could be passed from one individual to another by tiny germs, which we now call bacteria and viruses. This encouraged others to look more closely at these smallest of all living things. In doing so, scientists discovered that all life, from the most complicated to the simplest, seemed to be linked together at a basic level, in the structure of the cell itself.

☀ PASTEURIZATION

Pasteur's work led to a way of keeping milk fresh that is still used today. In the pasteurization process milk is heated for long enough to kill all the bacteria in it. The liquid can then be stored for some time with no hazard to health.

CELL DIVISION

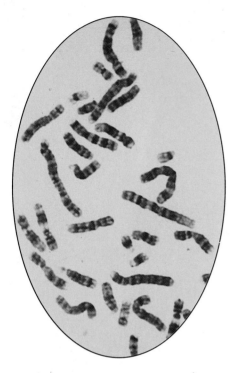

▲ A selection of human male chromosomes, shown magnified about 3,000 times. There is a total of 46 chromosomes in each of our cells.

Although Robert Brown discovered the nucleus in 1831, it was very difficult to see any detail inside a cell. When looked at through a microscope, cells appear to be almost transparent. Also, the biggest animal cells are no more than a tenth of a millimetre across, and most are much smaller. Anything that fitted inside this small space would have to be very tiny indeed. In 1842, Karl von Nägeli gave one of the first accurate descriptions of the way a cell divides. He saw the splitting of the nucleus in a dividing cell and the formation of tiny bodies in the nucleus before it divided.

With the invention of artificial dyes in the 1850s, biologists had a way of making the structures inside a cell show up more clearly. By the 1870s, cytologists, the name given to people who study cells, had discovered that different parts of the cell would take up different dyes. One of these people was Walther Flemming (1843-1905), a German anatomist. In 1879, while studying animal cells using the new techniques, he discovered a substance inside the nucleus that took up dye particularly well. He called the new substance chromatin, from a Greek word that means 'colour'.

Flemming was able to see how the chromatin behaved as the cell divided. First it formed into strands as division began. This was what von Nägeli had described over 30 years earlier. (In 1888 another German scientist, Wilhelm Waldeyer (1836-1921), gave the strands the name 'chromosomes', which means 'coloured bodies'.) Flemming saw that the chromosome strands were always present when the cell divided. Because of this, he gave cell division the name 'mitosis', from the Greek word for 'thread'. In the next stage in the division of the cell, the chromosomes doubled in number. Then they were pulled apart, half going to one end of the cell and half to the other. The cell then divided in two, leaving each new cell with the same number of chromosomes as the original cell. Finally, the chromosome strands disappeared as a new nucleus formed in each of the two new cells.

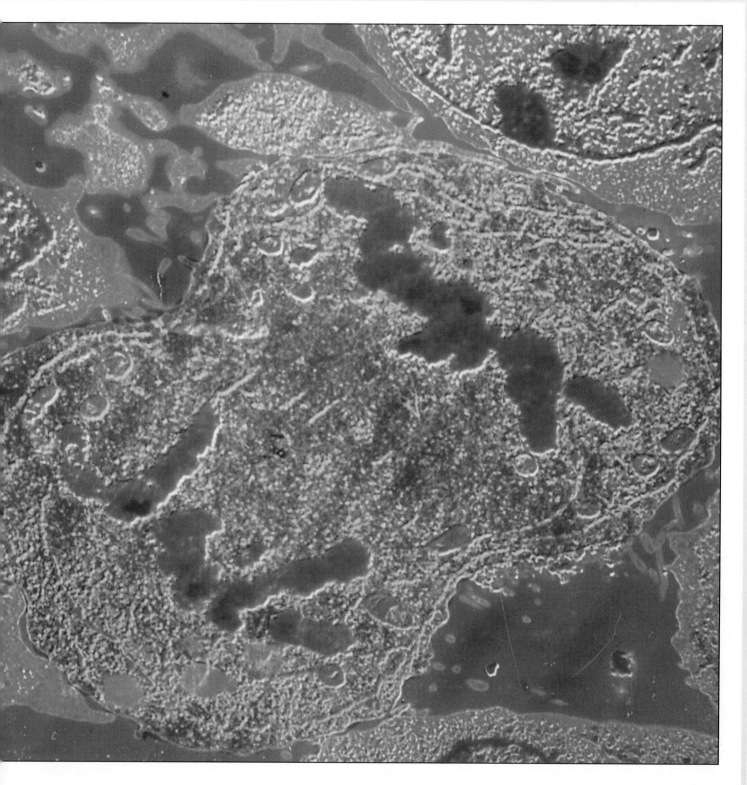

Flemming published his findings in a book called *Cell Substance, Nucleus and Cell Division* in 1882. Had he known of Gregor Mendel's work, he might have realized the importance of his discoveries in explaining Mendel's inheritance particles. Von Nägeli, who *did* read Mendel's findings, failed to see their significance.

▲ As a cell divides, the nucleus splits and small bodies form just before this happens. These are the chromosomes, which contain the blueprints of life for the generations to come.

MENDEL REDISCOVERED

In 1887, Edouard van Beneden (1846-1910), a Belgian cytologist, discovered that every cell in an organism has the same number of chromosomes, and that this number is the same for every member of a species. For example, in every healthy human cell there are 46 chromosomes. However, when the sex cells – the sperm and eggs – are formed, the number of chromosomes is halved. Human sperm and eggs each have only 23 chromosomes.

The German biologist August Weismann (1834-1914) tried to come up with a theory that could provide an explanation of heredity – the passing on of characteristics from one generation to the next – that tied in with what was known about the cell. In the 1880s, Weismann suggested that the chromosomes carried the hereditary material, and that this was passed from parents to their offspring. When fertilization took place, a sperm and egg joined together, uniting the chromosomes from two individuals. In this way, Weismann explained, the offspring would have the full number of chromosomes – half from its mother and half from its father. It was the combining of different chromosomes in this way that brought about the variations that were seen between individuals.

▼ Male sperm meets female egg in this electron microscope's view of the moment of fertilization. The magnification is some 50,000 times.

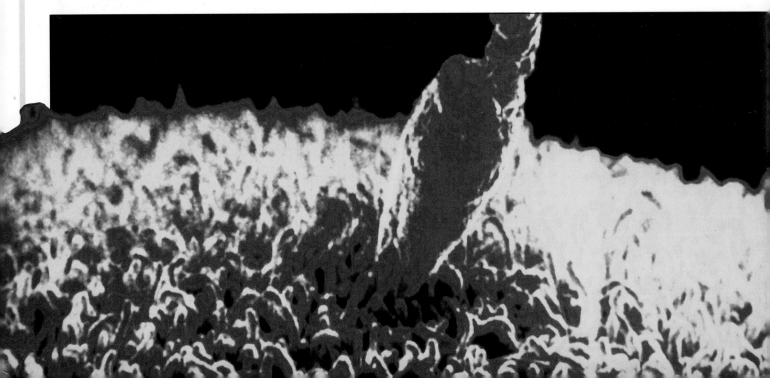

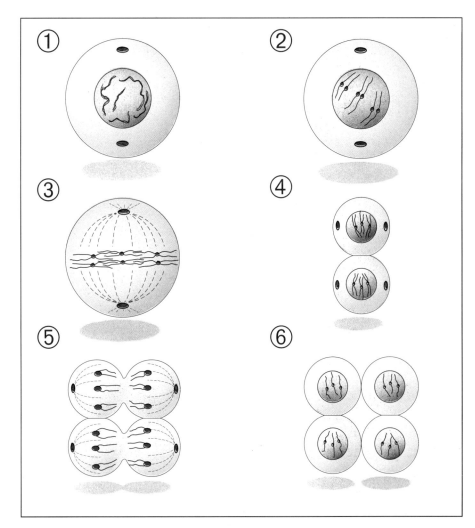

◄ **Meiosis is the special process through which male and female sex cells are formed. Just half the chromosomes are present, so that when male and female sex cells join, the new cell will have the correct number, half from each of the sexes.**

1 Small bodies called <u>centrioles</u> move either side of the central nucleus, inside which are the chromosomes.

2 Chromosomes become shorter and fatter. Then they pair up and divide into strands.

3 Chromosomes range across the cell's middle.

4 Cell divides.

5 Cells divide again.

6 Each cell now has half the chromosomes of the original cell. At fertilization, male and female cells fuse to provide a full set of chromosomes.

Hugo de Vries (1848-1935), a Dutch botanist, was also interested in how different characters were passed down the generations. He had read Darwin and was aware that evolution theory did not explain how variations came about between individuals of the same species. De Vries suggested that each characteristic was a distinct unit, so that each one could vary independently of the others, and could come together in different combinations. De Vries called these units 'pangenes'. In 1900, he was ready to publish his theory, but before doing so he looked carefully to see if anyone else had come up with a similar idea. To his surprise he found that someone had – Gregor Mendel.

For thirty-five years, Mendel's work had been unnoticed, published only in an obscure journal. De Vries rediscovered that work and announced it to the world, sixteen years after Mendel's death. The only credit de Vries claimed for himself was to say that his own work confirmed it.

☀ A REMARKABLE COINCIDENCE

Hugo de Vries was not the only one to find out that Mendel had got there first. Two other biologists working independently on plant hybridization at about the same time thought they had discovered the nature of heredity. The German Karl Correns and the Austrian Erich Tschermat von Seysenegg got the same results from their plant experiments as Mendel, and they all credited him as the great pioneer of genetics.

THE GENE

▲ August Weismann, whose work included proving the difference between inborn and acquired characteristics. One of his experiments included cutting the tails off mice to prove that despite this, their offspring would still be born with normal tails.

The hunt was on to establish where Mendel's particles of inheritance were to be found in the cell. Weismann's work suggested strongly that chromosomes were part of the solution. However, there are only twenty-three pairs of chromosomes in a human cell, yet humans have thousands of different characteristics. This had to mean that each chromosome carried thousands of Mendel's particles. In 1909, the Danish botanist Wilhelm Johannsen (1857-1927) suggested that the particles be called genes, from a Greek word that means 'to give birth to'.

In 1905, the English biologist William Bateson (1861-1926) discovered that the characters for pollen shape and flower colour were sometimes linked together, so that a plant with a particular shape of pollen was likely to have a particular flower colour, too. Some genes, it appeared, were linked and not inherited independently, as Mendel had supposed. Bateson also suggested a name for this new science of genes and heredity: genetics.

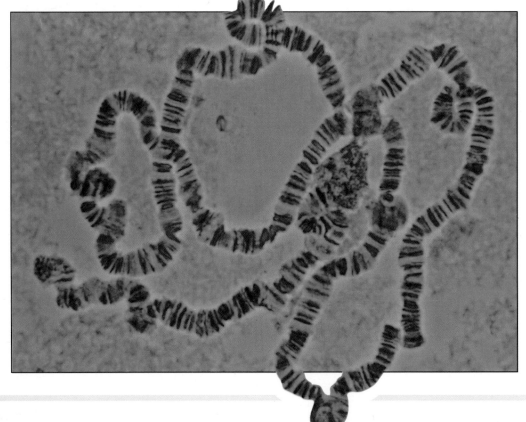

◀ The chromosomes of a fruit fly. They are very large, making them easy to study. The genes are contained in the dark bands. Each of these 'gene site' areas is known as a locus.

▲ Shown in closeup, one of the geneticist's favourite test creatures, the fruit fly. The creature's rapid breeding is ideal for fast checking between generations. Changes in the chromosomes (known as mutations) are common too, making for interesting studies.

The most important work in the new science was being done by the American geneticist Thomas Hunt Morgan (1866-1945). In 1907, he began a series of experiments at Columbia University, New York, using the fruit fly *Drosophila melanogaster*. Fruit flies were easy to keep, bred quickly and had only four pairs of chromosomes, so there was less material to examine.

The results of Morgan's experiments confirmed Mendel's laws in most cases. However, he found that whether or not the effect of certain genes showed up in the offspring depended on the sex of the parent that passed them on. For example, when white-eyed males were crossed with red-eyed females, all the offspring had red eyes. However, when white-eyed females were crossed with red-eyed males the female offspring had red eyes, while males had white eyes. White eyes were almost always found only in males.

Morgan believed that these results meant that the gene for eye colour was linked to the gene that determined the sex of the offspring. The explanation for this had to lie in the chromosomes. If linked genes were carried on the same chromosome, that would explain why they were inherited together. Morgan saw here a way to establish where the genes were to be found on the chromosomes.

☀ MALE OR FEMALE?

In complex animals, such as birds, reptiles and mammals, sex is determined by a pair of special sex chromosomes. In humans, women have two identical 'X' chromosomes, but men have one X and one 'Y' chromosome. Whether you are male or female depends simply on whether or not you have the Y chromosome.

MUTATIONS AND CHROMOSOME MAPS

▲ **Three years work went into Thomas Hunt Morgan's finding out where the genes were positioned on a fruit fly's chromosomes.**

The Dutchman Hugo de Vries put forward a new idea of evolution, suggesting that it progressed in a series of sudden jumps or changes. The name de Vries gave to these sudden changes was mutations. Every so often, an individual would appear that was different in some way from either of its parents. If the difference was one that gave an advantage, then it could be preserved and passed on to the following generations. Farmers had been aware of these changes for centuries. In 1791, for example, a mutant sheep with short legs was preserved because it couldn't jump over fences!

Thomas Hunt Morgan tried to cause mutations deliberately in his fruit flies and then follow them from generation to generation. He failed to do this, but did find enough examples of natural mutations to allow his work to continue. He was able to determine that many genes were inherited together, which would happen if they were on the same chromosome. However, sometimes there were exceptions, and the linked genes could be inherited separately.

◀ **Most mutations are small or unnoticeable and most do not breed true in future generations. Albinoism is a fairly common mutation, especially in domestic animals such as rabbits. This albino frog comes from the Rio Mazan valley in Andes Mountains of Ecuador, in South America.**

▲ Radiation from radioactive materials can badly damage living cells, and in the 1950s, much was made of the possible monsters that might result from the nuclear weapons (such as this hydrogen bomb) that would be used in a future war.

Morgan found that when the sex cells were forming, pairs of chromosomes exchanged part of their length with each other. This was called crossing over, an idea that had been suggested by F.A. Janssens in 1909. In 1911, Morgan put forward his chromosome theory of inheritance. In this, he said that the greater the length of chromosome that separated a pair of genes, the more likely it was that they might split up during crossover.

One of the assistants in Morgan's 'fly room' as it was called, was Alfred Sturtevant (1891-1970), a student at Columbia University. In 1911, he had a flash of inspiration. He realized that the distance between genes on a chromosome could be determined by measuring the frequency with which they crossed over. By making a careful study of how often a particular link was broken, Morgan and Sturtevant worked out where on the chromosome the genes could be found. Soon, they were able to draw up the first fruit fly 'chromosome map', showing the positions of five genes that were linked to the sex gene. Morgan had taken Mendel and genetics just about as far as it could go. There was no doubt now where genes were located. The next question to answer was, how did they work?

GENES AND DNA

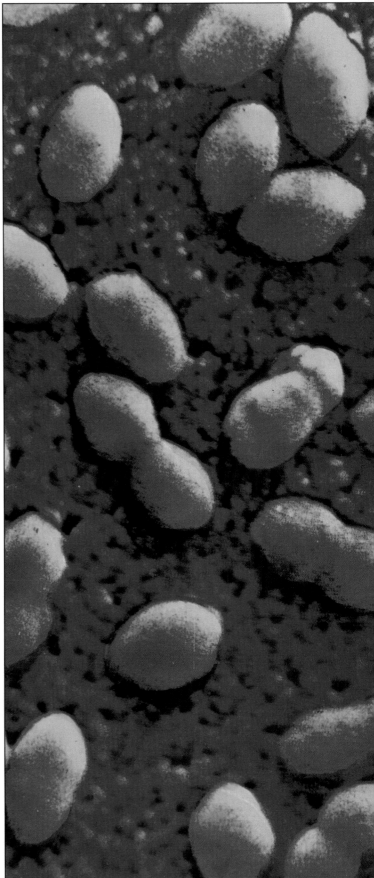

At the beginning of the twentieth century the cell was being studied in greater and greater detail. Scientists such as the Russian-born Phoebus Levene (1869-1940) studied the way the cell worked on the chemical level. Among other things he studied nucleic acids. These were substances found in the nucleus of cells, which were unlike any other natural chemical known at the time.

In 1909, Levene was able to show that one of the nucleic acids contained a type of sugar called ribose, so this became ribonucleic acid, or RNA for short. Twenty years later, in 1929, he found that other nucleic acids contained a sugar called deoxyribose. This became known as deoxyribonucleic acid, or DNA. These were soon to be revealed as the most important chemicals in life. No other nucleic acids have been found. In 1944, Oswald Avery (1877-1955), a Canadian physician, was studying two of the bacteria that caused pneumonia. One type of bacterium appeared smooth on the outside: this was called an S-type. The other was rough looking – the R-type. Avery discovered that by mixing an extract of dead S-types with live R-types, he ended up with live S-type bacteria.

▶ Pneumonia bacteria, similar to the ones that Oswald Avery studied during his researches. These are shown enlarged about 20,000 times.

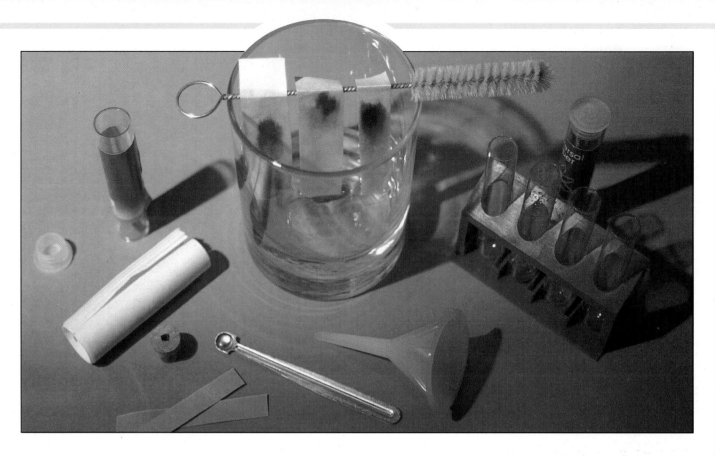

It seemed that there was something in the S extract that made the R-type bacteria change into the S types. Avery discovered that the chemical responsible was DNA. He suggested that DNA was the chemical at the heart of heredity. If the structure of DNA could be determined it might lead to the secret of the genes. It was a challenge that was soon taken up. In the same year that Avery carried out his research, the English biochemists Richard Synge (b.1914) and Archer Martin (b.1910) developed the technique of chromatography. This made it possible to separate chemical mixtures into their constituent molecules.

Erwin Chargaff (b.1905), a Czech-born biochemist, used chromatography to study DNA. The structure of the individual parts of the DNA molecule, called nucleotides, was fairly well known, but not how they all fitted together. There were four nucleotides, each with a different chemical base – these were adenine, thymine, cytosine and guanine. Chargaff showed that the bases were always found in pairs. There were the same number of cytosine bases as guanine, and the same number of adenine as thymine bases. Also, the number of adenine and guanine bases equalled the number of cytosine and thymine bases. These were vital clues to the structure of DNA.

▲ Chromatography is used to separate out different chemicals. This simple chemistry set example uses paper as a medium, with drops of ink (each a mixture of different dyes) being placed on strips of filter paper, suspended over a glass. Liquid is added (here it is water). As the water soaks up into the paper, the blots gradually spread out. Heavier dyes stay near the bottom, lighter ones float up with the water until all are separated.

THE DOUBLE HELIX

▲ **James Watson (left) and Francis Crick, photographed in 1953 with their first model of the DNA molecule.**

Often in science, as in other walks of life, all the clues are available to solve a mystery. All that is needed is that they be put together. By 1951, English scientists Maurice Wilkins (b.1916) and Rosalind Franklin (1920-58), working at King's College, London, had used a new technique, called X-ray crystallography, in an attempt to work out the structure of the DNA molecule. Their evidence seemed to point towards it having a spiral or helical shape. Franklin was not convinced this was right, however, and wanted to find further evidence to confirm it.

Meanwhile, in Cambridge, England, the American geneticist James Watson (b.1928) and the English physicist Francis Crick (b.1916), were busy gathering together everything that was known about the structure of DNA. They had been working together since 1951, when Watson had suggested that the structure of DNA would show how it worked in the genes. In autumn that year, Watson had attended a lecture given by Franklin and tried to used what he learned there to build a model of DNA. Unfortunately, he misunderstood what he had heard and the result was a failure.

▶ **The DNA in human cells is made up of about 3 billion pairs of bases, divided up among the 46 chromosomes. If all the DNA in a single cell could be stretched out in a single line, it would extend to a length of about one metre. The genes actually take up only about two per cent of the DNA. The rest does not appear to carry a working code.**

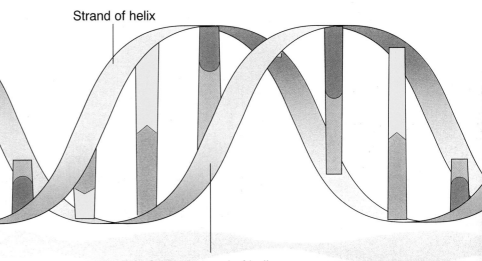

Strand of helix

Second strand of helix

In 1952, Watson and Crick met Chargaff and he told them what he had learned concerning the pairing of the bases in the DNA molecule. Watson again tried to make a model, using Chargaff's findings and those of the Scottish chemist Alexander Todd (b.1907), who had shown that the nucleic acids contained sugar and phosphate groups. The chemical structure of the bases was now well determined and Watson used simple cardboard cutouts to represent their shapes. He noticed that an adenine-thymine pair was the same shape as a guanine-cytosine pair. If every cytosine was paired with a guanine and every thymine with an adenine, it would explain why Chargaff had found the bases in equal numbers.

Crick took over the model building in February 1953. By the end of the month they had the answer. Their model revealed the DNA molecule as a double helix – two spiralling backbones of sugar-phosphate molecules twisted around the outside of the molecule. Attached on the inside, like the steps in a spiral ladder, were the base pairs. It incorporated all the known facts about DNA and confirmed that Franklin and Wilkins' X-ray patterns were right. As Watson and Crick themselves noted, the secret of DNA and genes lay in how the pairing of the bases provided a way of making accurate copies of the DNA molecule.

THE NOBEL PRIZE

The Nobel prize is a group of awards given to achievers in physics, chemistry, medicine, physiology, literature and work for peace. In 1962 James Watson, Francis Crick and Maurice Wilkins were jointly awarded the Nobel Prize for medicine for their pioneering work in discovering the structure of DNA.

Absent from these awards was a prize for Rosalind Franklin, who died of cancer in 1958. The Nobel Prize committee does not make awards posthumously, so Franklin did not get the official recognition that her work deserved, although Watson described it as 'superb'.

▼ **DNA has two long strands, coiled around each other in a spiral known as a double helix. Just as a vast number of words can be made from the letters of the alphabet, so DNA can generate huge quantities of genetic instructions from its simple code. A DNA molecule may vary in size from a mere 100,000 atoms or so to a more complex 10 million-plus.**

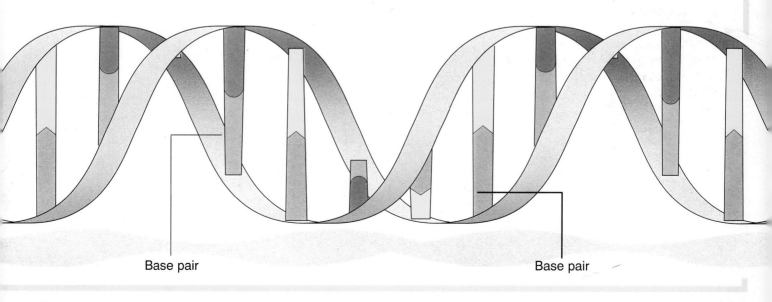

Base pair

Base pair

DNA AT WORK

Crick and Watson put forward the idea that if the two strands of the DNA molecule were separated, each strand could act as a template, or copying guide. This would be used to form a new strand, identical in every way to its original partner. There would then be two DNA molecules, each of which had an original strand and a new one. This tied in with the doubling of the chromosomes that was seen when cells divided.

Crick and Watson also said that the order the chemical bases came on the DNA molecule could represent a code, to be translated by the cell and used as a guide for making proteins. These are complex molecules found in all living things, performing a multitude of tasks, including carrying substances around the body, defending against infection, allowing muscles to move and nerves to carry messages. A very important job, done by proteins called enzymes, is controlling the chemical reactions that go on in living cells.

▶ **When copying itself, the strands of the DNA molecule come slowly apart. The base pairs separate and new bases join the remaining ones on the strands. The result is a pair of DNA molecules identical to the original one.**

By 1961, after Watson had returned to America, Crick had shown that each three-base group on a DNA strand is the code for one of the chemical building blocks from which proteins are made. Proteins themselves are composed of long strings of simpler molecules called amino acids. The way these are arranged determines the three-dimensional shape of the protein and the job it does. To make new proteins, a small section of DNA unwinds and a single strand of RNA – the other nucleic acid discovered by Levene – is formed by using one of the unwound strands as a template. This is called messenger RNA.

The messenger RNA leaves the cell's nucleus and travels to other parts in the cell called ribosomes. A ribosome 'reads' the code carried on the RNA and amino acids are assembled to form protein, according to the order the bases come on the RNA. This is the genetic code at work. DNA controls which proteins are made; proteins control the organism's activities.

GENETIC MISPRINTS

The genetic code is common to all forms of life. A gene can be thought of as a section of a single strand of DNA. It may, for example, be responsible for making an enzyme protein that controls the manufacture of a pigment that gives a flower its colour. Through chains such as this the DNA guides the appearance of the organism. Its environment can also influence its appearance to some extent.

Sometimes mistakes are made when DNA is being copied. Just as a misprint in a book can change the meaning of a sentence, so a copying mistake in DNA changes the nature of the organism that carries it. If mistakes are made when sex cells are being formed, the offspring may be different from its parents because of the alteration. Most mistakes, or genetic mutations, are harmful and offspring may not even survive until birth. But sometimes such mutations provide advantages and it is these that provide fuel for evolution.

☀ THE STUFF OF EVOLUTION

Mutations are not always bad news. In many cases, they help species produce more offspring or compete more successfully for food. Scientists describe such mutations as 'selectively advantageous'. In fact, mutations are the reason why there is so much variation among living things. It is the raw material on which natural selection can do its work. And this carries on today, with unpredictable results for the future of life on Earth.

◀ **Georges Buffon defined a species as a group of living things that can breed with each other, but not with other species. Migrating species can move apart so far that eventually, small accumulated changes add up so much that a new species is the result. These two species of gull once had a common ancestor that originated in Siberia.**

If an animal is faster, for example, it might have a better chance of surviving long enough to reproduce. The advantages the mutation has given it would then be passed on to the offspring. Evolution works through mutation as de Vries suggested, and accidental errors in copying provide the range of variations within species that Darwin and Wallace had seen. Natural selection then gets to work, usually weeding out variations that do harm and preserving ones that give an advantage. New species come about through DNA mutations being preserved and passed on, and also by the crossing-over process during meiosis.

The question scientists are now asking is: why wait for random mutations to produce new life forms? Could it be done more efficiently in the laboratory?

GENETICS TODAY

The study of genetics has transformed research in the life sciences – areas such as biology, zoology, botany and biophysics. Among the questions that scientists are asking today are: can we cure inherited diseases? Can we understand biochemical mechanisms in the body, especially the proteins that govern all living processes?

In fact, much genetics work is done in these fields. Hopefully, with better understanding of genes and how they work we will be able to cure many diseases, including cancer, HIV infections and heart disease. In genetic engineering (the term now used to describe the geneticists' work) there have been many important advances. One came in 1973 when American geneticists Stanley Cohen (b.1935) and Herbert Boyer (b.1936) used enzymes to slide out sections of DNA from different bacteria, then recombined them to form a completely new strand, called 'recombinant DNA'. With this technology, Boyer showed how bacteria could be used to make human proteins such as insulin, which is used to treat diabetes.

◄ A laboratory mouse was the first animal to be patented as a genetically-engineered creature.

DNA PROFILING

This is a technique that was developed in 1984 by British scientist Alec Jeffreys (b.1950) at Leicester University. DNA profiling identifies and matches short sequences of DNA that are unique to each of us. As a result, DNA profiling can be used to identify an individual from traces of blood or skin cells found at the scene of a crime. It must be used with traditional methods of crime detection though, as the system is not infallible. In a country the size of the USA for example, some 300 people might have DNA profiles that look very similar.

Many more advances have followed in the years since Cohen and Boyer's work. In 1980 the US Supreme Court granted the world's first-ever patent on a living organism – it was a genetically-engineered oil-eating bacterium intended to play a big part in clearing up oil spills from damaged tankers and oil rigs. New animals have also been created in the laboratory. In 1988 a new strain of mouse was patented in the USA. The creature had been been developed in the laboratory especially for use by scientists working on cancer research projects.

In agriculture there have been many advances, too. Food crops have been genetically improved, with genes added to make the plants more resistant to weed killers, so that weeds can be sprayed and destroyed without farmers having to worry about damaging the crop itself. Crops have also been made more resistant to leaf-eating insects and other creatures that attack plants in the fields.

One of the biggest genetic projects now underway in laboratories all over the world – and often called the 1990s equivalent of the Apollo moon-flight programme – is the mapping of the entire human 'genome', all the genetic information carried in our chromosomes. The project is a massive job, as the human DNA double helix has billions of rungs on it, but the rewards in the medical world are likely to be huge, not least in the area of checking for genetic damage in people's chromosomes. By the early years of the twenty-first century there could be tests for a thousand or more inherited diseases, with cures or relief available for many of them. The most important use of genetics today is in designing new drugs to try and combat diseases such as Aids, Parkinson's, Alzheimer's and cancer.

▲ DNA 'fingerprints' are made by putting a marker chemical on to DNA fragments and placing them in a gel. The fragments leave dark marks on a piece of X-ray film when an electric current is passed through the gel. The resulting stripes, which look rather like a blurred version of a shopping bar-code, can then be examined to see if they are similar to a test sample taken from a criminal suspect.

INTO THE FUTURE

Genetic engineering is going to be one of the major technologies of the next century, with the results visible everywhere, from kitchen table to motorway. Today's 'Flavr-Savr' tomatoes are a taste of things to come. Tomatoes normally ripen on the vine, but spoil on their way to market if picked at the moment of ripening. At present, they have to be picked early to prevent spoiling, with resulting loss of flavour. The Flavr Savr has been bio-engineered with an extra gene to slow the natural ripening process – it can be left to ripen naturally because it remains firm for much longer after it has been picked.

Other gene alterations sound more like astounding stories from the worlds of science-fiction. For example, many Antarctic fish thrive in icy waters because they have special proteins that act as a natural 'anti-freeze' – and it may be possible to protect soft fruits such as strawberries and raspberries from frost damage by transferring genes from these fish!

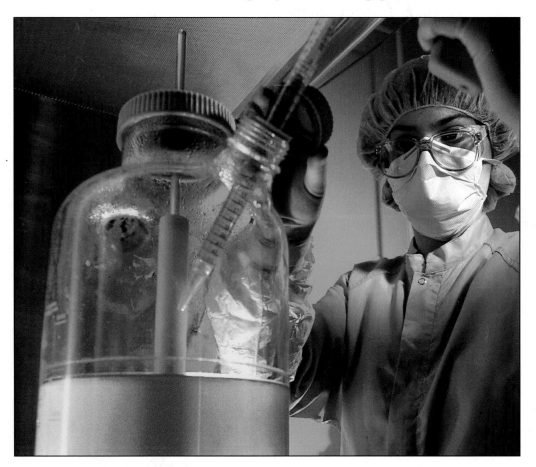

◄ Genetically identical antibodies being made for the production of a life-saving vaccine. Genetic disorders that are high on the list for researchers include Alzheimer's disease, which causes memory loss. Parkinson's disease affects the nervous system, while cystic fibrosis produces mucus that makes breathing difficult.

Supplies of crude oil are likely to run short in the next century. Currently, about half the world's supply of oil is used for transport, with the rest used for making such things as plastics, drugs, pesticides and paints. Genetic engineering may come to the rescue, by creating new breeds of plant that bypass drilling rigs and refineries entirely. Special-breed potatoes may one day produce a type of plastic instead of starch. And when you fill up your clean-burning future car, it may be with petrol produced by gene-engineered crops such as oil-seed rape and sunflowers. And of course, fields of such crops are more friendly to the environment than oil-rigs and chemical works.

So far, plants have been the main area for genetic engineering as they are easier to work with than animals. However, scientists have introduced human genes into the fertilized eggs of pigs, mice, sheep and cows. The target is to perfect a technique known as 'pharming' – getting the animals to make human proteins for medical use. The most likely first products of the new pharming industry are proteins to treat problems such as blood disorders, cystic fibrosis and blood clots. Once production is running on a commercial scale, pharming should be much cheaper than present-day methods and safer too, with little risk of contamination, as the proteins are planned to be supplied in the animals' milk.

RIGHT OR WRONG?

Genetic engineering seems wonderful, but there are many rights and wrongs to this new technology. For example, genetic screening for defects creates a dilemma – should you tell a relative that he or she has something wrong? Should you have a family if you have a defect that may show up in your children?

Many companies demand big money for the rights to use their newly-created genes. This is understandable, as research costs are very high, but it encourages crime, which came to genetics in 1994, when two men were charged with 'gene stealing'. They were accused of taking cells from an American laboratory, then demanding $300,000 for them for outlaw production. The cells contained a hormone gene for treating kidney defects.

TIMELINE OF ADVANCE

H ere are some of the people whose science discoveries have helped to bring about today's world.

Francesco Redi Italian biologist (1626-97). Showed that flies did not appear spontaneously from rotting meat, but from eggs laid on the meat by other flies.

Robert Hooke English scientist (1635-1703). Produced the book *Micrographia*. Introduced 'cell' to describe the basic unit of life.

Carl Linnaeus Swedish botanist (1707-78). Worked out a system for classifying living things, published in his book *Systema Naturae*.

Georges Buffon French naturalist (1707-88). Suggested that changes appeared in a species if it had to live in a new environment.

Charles Bonnet Swiss naturalist (1720-93). Speculated that every female contained the 'germs' or 'seeds' that would grow into every generation to follow. Also believed that new species appeared following world-wide catastrophes.

Erasmus Darwin English scientist (1731-1802). Grandfather of Charles Darwin, he put forward a theory that species could adapt to

their surroundings and that this could bring about the evolution of new species.

Jean-Baptiste Lamarck French naturalist (1744-1829). His theory suggests that characteristics acquired during an animal's lifetime could be passed on to its offspring.

Robert Brown Scottish botanist (1773-1858). Discovered the nucleus of the cell. Brown helped to classify some of the plants that Charles Darwin brought back from the *Beagle* voyage.

Carl von Siebold German biologist (1804-85). Established that some organisms were composed of just a single cell. He called this new group of living things the *Protozoa*.

Matthias Schleiden German botanist (1804-81). Recognised that all plants consisted of cells that contain a nucleus.

Charles Darwin English naturalist (1809-82). Became world-famous for his theory of evolution by natural selection, suggesting that nature selected individuals and species best fitted to their environment.

Theodor Schwann German biologist (1810-82). One of the first to realize that all animal tissues

consisted of cells that contained a nucleus just as Schleiden had observed in plants.

Karl von Nägeli Swiss botanist (1817-1891). Possibly the first to see chromosomes during cell division. However, he is usually remembered as the man who dismissed the work of Mendel, perhaps setting back the science of genetics by several decades.

Rudolph Virchow German doctor (1821-1902). Helped to develop cell theory by showing that all cells developed from other cells. He summarized this in the phrase, 'all cells arise from cells'.

Gregor Mendel Austrian monk and botanist (1822-84). Often referred to as the 'father of genetics'. Mendel's experiments with heredity in peas, carried out in his monastery garden, established his laws of inheritance.

Louis Pasteur French chemist (1822–95) Proved once and for all that spontaneous generation was false. Also developed the germ theory of disease, which revolutionized medicine.

Alfred Russel Wallace English naturalist (1823-1913). Came up with an evolution theory based on natural selection that was almost identical to that of Charles Darwin. The two men agreed to a joint presentation of their ideas in 1858.

August Weismann German biologist (1834-1914). First suggested that the chromosomes in a cell passed on characteristics to the next generation.

Wilhelm Waldeyer German anatomist (1836-1921). Introduced the word 'chromosome'. Also did important work on the nervous system, showing it to be made up of separate cells.

Walther Flemming German anatomist (1843-1905). Using new dyeing techniques, he succeeded in showing cells at various stages of division, clearly revealing the thread-like chromosomes.

Edouard van Beneden Belgian cytologist (1846-1910). Discovered that the number of chromosomes in a cell is constant for each species, and that the number is halved when the sex cells are formed.

Hugo de Vries Dutch botanist (1848-1935). Theorized that evolution proceeds in a series of sudden changes or mutations. Rediscovered the work of Mendel and ensured its wide acceptance.

Wilhelm Johannsen Danish botanist (1857-1927). Suggested the name 'gene'.

William Bateson English biologist (1861-1926). His plant-breeding experiments showed that some characteristics were linked, and not inherited independently.

Thomas Hunt Morgan American geneticist (1866-1945). Worked with mutations in fruit flies and produced the first chromosome map showing where the genes for various characteristics were located.

Phoebus Levene Russian-American chemist (1869-1940). Worked on the structure of nucleic acid and discovered that there were two forms: ribonucleic acid (RNA) and deoxyribonucleic acid (DNA).

Oswald Avery Canadian doctor (1877-1955). Discovered that DNA was the chemical that held genetic material of cells. Previously, DNA had been thought to be an unimportant part of the nucleus.

Alfred Sturtevant American geneticist (1891-1970). Worked in Thomas Hunt Morgan's 'fly room' at Columbia University while a student. Realized that the distance between genes on a chromosome could be calculated by observing how often they were inherited separately.

Erwin Chargaff Czech-born biochemist (b.1905). Discovered that the bases, which are important components of DNA, fell into pairs.

Alexander Todd Scottish chemist (b.1907). Studied the chemistry of various compounds found in living organisms, including the nucleic acids, DNA and RNA. Awarded the Nobel prize for chemistry in 1957.

Archer Martin English biochemist (b.1910). With Richard Synge, he developed chromatography, a method of separating out the components of mixtures.

Francis Crick English physicist (b.1916). With American James Watson (b.1928), he discovered the structure of the DNA molecule in 1953. Crick went on to show how the genetic code held in DNA operates in the cell.

Maurice Wilkins New Zealand-born British physicist (b.1916). Worked with Rosalind Franklin to try to determine the structure of DNA. Their work was vital to Watson and Crick's ultimate success in building a model of DNA.

Rosalind Franklin English chemist (1920-58). Her X-ray pictures of DNA pointed towards its helical structure and provided Watson and Crick with several vital clues in their DNA research.

Stanley Cohen American molecular geneticist (b.1935). Worked with Boyer on techniques for recombining fragments of DNA.

Herbert Boyer American geneticist (b. 1936). Showed that sections of DNA from different bacteria could be spliced together and so opened the way for genetic engineering, a new technology the techniques of which are widely used today.

GLOSSARY/1

A ready-reference guide to many of the terms used in this book.

Acquired characteristics Features that an organism develops as a result of its own efforts. A body builder's muscles are an example of acquired characteristics. At the beginning of the 19th century Jean-Baptiste Lamarck proposed that the passing on of these characteristics from generation to generation could explain the process of evolution.

Adaptation Feature that gives a plant or animal a better chance of survival. For example, the streamlined body shape of dolphins, penguins and many fish is an adaptation for fast swimming.

Amino acids The chemical building blocks from which proteins are made. Plants and many micro-organisms can make all the amino acids they need from simple compounds taken from their surroundings. Animals can make some amino acids but others, called the essential amino acids, must be found in their food.

Ancestor An individual from whom others are descended and to whom a direct line can be traced through parents, grandparents, great grandparents, and so on.

Base One of the parts of the DNA molecule. There are four types of base in DNA. The specific pairing of the bases holds the two strands of DNA together and allows accurate copying of the genetic code.

Cell The basic unit that makes up living organisms. Cells can exist singly as independent life forms, such as bacteria, or they may group together in more complicated organisms, such as ourselves. Each cell has a central nucleus containing its DNA, surrounded by a jelly-like substance. The cell is bounded by a thin membrane, which in plants, fungi and bacteria is also surrounded by a rigid cell wall.

Chromatin A substance, found in the nucleus of a cell, of which chromosomes are composed. It contains proteins, DNA and RNA.

Chromatography A technique for separating out mixtures of gases, liquids or dissolved substances.

Chromosomes Coiled structures found in the nucleus of a cell. Chromosomes carry the genes that determine the characteristics of the organism. They only become visible in the nucleus when the cell is dividing. The number of chromosomes in a cell is always the same for each species. Bacteria do not have a nucleus and have only a

▲ **Dolphins, penguins and many fish have evolved similar-shaped bodies, even though dolphins are mammals, penguins birds. This is** a case of parallel evolution, in which similar requirements, speed under water, have provided the same streamlined answer.

single chromosome; maize cells have ten pairs and each human cell contains 23 pairs of chromosomes.

Crossing over This refers to the exchange of genetic material between two chromosomes during cell division when the sex cells are being formed. Crossing over changes the pattern of genes on the chromosomes.

Cytology The study of cells and their functions. Scientists working in this field are called cytologists.

DNA Stands for *deoxyribonucleic acid*, the genetic material of all living organisms, apart from viruses. DNA consists of two long chains of nucleotides joined together and coiled into a shape something like a twisting ladder – the double helix. When a cell divides, its DNA unravels into two separate strands. Each strand rebuilds its opposite half from free nucleotides in the cell nucleus. In this way two identical copies of the original DNA are formed. Cells manufacture proteins using the information encoded in the DNA.

DNA profiling Often called genetic fingerprinting, this is a technique, sometimes employed by police, for identifying people through the unique structure of their DNA.

Evolution The gradual development of new plant and animal species over many millions of years. It is now widely believed that the changes come about

◄ **Eye colour can be a clue to your genetic heritage. These sisters both have brown-eyed parents, yet blue-eye recessive genes from the grandparents' generation has provided the younger child with blue eyes.**

through natural selection, a view first put forward by Charles Darwin.

Fertilization The joining together of male and female sex cells, for example sperm and eggs in animals, during reproduction. Only after this has happened can a new organism begin to develop.

First filial generation The first generation born following the parents' generation, i.e. sons and daughters. Grandchildren would be the second filial generation.

Gene The basic unit of heredity. Genes are composed of DNA and form part of a chromosome. They determine the particular characteristics an organism inherits from its parents, such as eye colour or seed shape.

Genetic engineering Changing the genetic make-up of an organism by artificial means, such as splicing genes from another organism into

its DNA. Genetic engineering may be used to produce vaccines to combat disease, or new strains of plants and animals that combine desirable characteristics from two or more different types. Genetic engineers often makes use of gene libraries, collections of DNA fragments made by breaking up the DNA of an organism by chemical and physical means.

Genetics The study of the ways in which characteristics are passed from generation to generation.

Genome The complete collection of genes in a cell.

Heredity The passing on of characteristics from one generation to the next.

Hybrid Offspring produced by crossing parents with distinct genetic differences, such as mating a horse and a donkey to produce a mule.

Linkage In genetics, the likelihood of a number of characteristics being inherited together because the genes responsible for those characteristics are located close together on the same chromosome.

Messenger RNA RNA stands for *ribonucleic acid*, a complex compound similar to DNA. This messenger RNA plays a crucial role in the formation of proteins. It is formed in the cell nucleus by copying a part of the cell's DNA that has the code for a protein. It takes this code from the nucleus to the ribosomes, where specific amino acid molecules are brought into the correct positions along the messenger RNA and linked together, building up the protein.

Mitosis Cell division, in which the original cell divides to form two new daughter cells, each containing the same number and kind of chromosomes as the mother cell. This is the way in which body cells normally divide.

Mutation A random change in the genetic material of a cell that may cause it to look or behave differently from a normal cell. An organism affected by mutation is called a mutant. Most mutations are harmful, but any beneficial changes may be transmitted down the generations, ultimately leading to evolution of the species.

Natural selection In his book *On the Origin of Species* (1859), Charles Darwin put forward the idea that those offspring of a plant or animal best suited to their environment have the best chance of survival. The survivors pass on their advantages to their offspring and this process results in a gradual change, eventually leading to a new species. Present-day species are all thought to have evolved through natural selection.

Nucleotides A group of chemical compounds that combine to form the nucleic acids, DNA and RNA. Each nucleotide has the same basic composition, consisting of a sugar molecule (deoxyribose in DNA and ribose in RNA) linked to a phosphate group attached to which is a compound called a base. There

◄ Living things have adapted and evolved to fill every niche on our planet. But adaptation has its limits and mass extinctions have happened many times in the past. The best known is that of the dinosaurs (a triceratops is shown here), some 65 million years ago. The reason why they disappeared is not known, but some scientists speculate that a massive comet may have hit the Earth, causing massive climate changes. This could have had disastrous effects, including loss of food and changed habitats.

are five bases – adenine, guanine, cytosine, thymine, and uracil, which is found only in RNA. The sugar-phosphate groups link together in chains, forming the 'backbone' of the DNA and RNA molecules. The bases link across the middle of the DNA in a very specific way – adenine pairs with thymine and guanine pairs with cytosine.

Nucleus Structure within a cell that contains its genetic material in the form of DNA. The nucleus is the cell's control centre.

Proteins A large group of organic compounds that is found in all living things. Proteins are made from long chains of amino acids, linked together in a sequence that is coded in DNA. A living organism may contain more than 100,000 different kinds of protein, all with a task to perform.

Recombination The reshuffling of genetic material that occurs when sex cells are formed. This results in offspring having combinations of characteristics that are different from those of their parents. Genetic engineering techniques can bring about recombination artificially.

Ribosomes Small spherical bodies found in a living cell, made of proteins and a form of RNA. This is where proteins are built up by RNA.

RNA Stands for *r*ibo*n*ucleic *a*cid. It is involved in making proteins in the cell (*see* Messenger RNA).

Sex cells Cells, such as sperm and eggs, that contain only half the usual number of chromosomes. When fertilization takes place, male and female sex cells join together to form an embryo that can develop into a new individual.

Species One of the categories used in classifying living things. Members of a species are basically alike and are capable of breeding among themselves and producing fertile young. Human beings, Californian condors and brown bears are all examples of species. Some different species, such as donkeys and horses, can mate successfully, but the offspring – in this case, a mule – cannot reproduce.

Spontaneous generation A belief once held that living things could arise from non-living materials.

X-ray crystallography A technique for studying the structure of crystals by using X-rays. A beam of X-rays is pointed at the crystal and the way they scatter is recorded on a photographic plate.

GOING FURTHER

Books There are several books on genetics and evolution. Try these for starters:
Breakthrough: Genetics by Tony Hooper, Simon & Schuster Young Books 1992
Eyewitness Science: Evolution by Linda Gamlin Dorling Kindersley 1993
Charles Darwin by Anna Sproule, Exley Publications
Genetic Engineering by Jenny Bryan, Wayland 1995
Magazines There are a number of general interest science magazines which often include articles on genetics. Try the following:
New Scientist A fascinating publication read by many scientists. Available in most libraries. Weekly.
Focus Large and small articles on a wide range of

subjects. Contains fascinating information with lots of pictures and diagrams. Monthly.
National Geographic Available on subscription, or for sale in better newsagents. Superb natural history articles are its strength. Monthly.
Places The best and most famous museum in Britain about the life sciences is London's *Natural History Museum*, which has a huge range of exciting displays. A branch of the museum in Tring, Hertfordshire, includes a spectacular display of creatures from all over the world that were found and stuffed by Victorian collectors. Darwin's home, *Down House* in the village of Downe, Kent, has many items to do with his life and journeys.

INDEX